Use of Multi-Node Wells in the Groundwater-Management Process of MODFLOW–2005 (GWM–2005)

By David P. Ahlfeld and Paul M. Barlow

Chapter 47 of
Section A, Groundwater
Book 6, Modeling Techniques

Groundwater Resources Program

Techniques and Methods 6–A47

U.S. Department of the Interior
U.S. Geological Survey

U.S. Department of the Interior
SALLY JEWELL, Secretary

U.S. Geological Survey
Suzette M. Kimball, Acting Director

U.S. Geological Survey, Reston, Virginia: 2013

For more information on the USGS—the Federal source for science about the Earth, its natural and living resources, natural hazards, and the environment, visit http://www.usgs.gov or call 1–888–ASK–USGS.

For an overview of USGS information products, including maps, imagery, and publications, visit http://www.usgs.gov/pubprod

To order this and other USGS information products, visit http://store.usgs.gov

Suggested citation:
Ahlfeld, D.P., and Barlow, P.M., 2013, Use of multi-node wells in the Groundwater-Management Process of MODFLOW–2005 (GWM–2005): U.S. Geological Survey Techniques and Methods, book 6, chap. A47, 26 p., http://pubs.usgs.gov/tm/06/a47/.

Preface

This report describes a new capability of the Groundwater-Management (GWM) Process in the 2005 version of the U.S. Geological Survey (USGS) modular three-dimensional groundwater model, MODFLOW–2005. The new capability allows the use of multi-node wells simulated by the MNW2 Package of MODFLOW to be part of the groundwater-management formulation. Previous versions of the GWM Process allowed the use of multi-node wells as unmanaged wells in the groundwater-flow process model but not as managed wells in the management formulation. The performance of the program has been tested in a variety of applications, some of which are documented in this report. Future applications, however, might reveal errors that were not detected in the test simulations. Users are requested to notify the USGS of any errors found in this report or the computer program.

Although this computer program has been written and used by the USGS, no warranty, expressed or implied, is made by the USGS or the U.S. Government as to the accuracy and functionality of the program and related program material, nor shall the fact of distribution constitute any such warranty, and no responsibility is assumed by the USGS in connection therewith. GWM–2005, MODFLOW–2005, and other groundwater programs are available online from the USGS at the following address: http://water.usgs.gov/software/lists/groundwater/.

Contents

Preface iii
Abstract 1
Introduction 1
Multi-Node Wells in Management Formulations 2
Multi-Node Wells Defined as Flow-Rate Decision Variables 2
Defining the Number and Locations of Managed Multi-Node Wells 3
Head-Loss Terms and Partial Penetration of Well Screens 3
Constraining Pumping Rates at Managed Multi-Node Wells 4
Unsupported Options of the MNW2 Package for Managed Multi-Node Wells 5
Management of Water Levels at Multi-Node Wells 5
Sample Problem: MNW–Supply 5
Impact of Multi-Node Wells on the GWM Algorithm 14
Summary 16
Acknowledgments 17
References Cited 17
Appendix 1. Input Instructions for Use of MNW-Type Wells in a GWM Formulation 19

Figures

1. Schematic diagrams of the model grid for the MNW–SUPPLY sample problem showing grid-cell locations of *A*, managed and unmanaged wells and *B*, specified-head constraints 6
2. Decision-variables (**DECVAR**) file for the MNW–SUPPLY sample problem 9
3. Decision-variable constraints (**VARCON**) file for the MNW–SUPPLY sample problem 10
4. Listings of the *A*, head-constraint (**HEDCON**), *B*, state-variables (**STAVAR**), and *C*, summation-constraint (**SUMCON**) files for the MNW–SUPPLY sample problem 11
5. Selected output from the GWM **OUT** file for the optimal solution of the MNW–SUPPLY sample problem 12
6. Selected output from the MODFLOW **LIST** file for the end of the second stress period and optimal pumping rates for the MNW–SUPPLY sample problem 13
7. Graphs showing groundwater-head responses at the end of the fourth stress period to pumping at well $Qm1$ for MNW–SUPPLY sample problem with input-variable LOSSTYPE equal to *A*, SKIN and *B*, GENERAL 15

Table

1. Characteristics of managed wells in the MNW–SUPPLY sample problem 8

Conversion Factors and Abbreviations

Multiply	By	To obtain
	Length	
foot (ft)	0.3048	meter (m)
	Volume	
cubic foot (ft^3)	0.02832	cubic meter (m^3)
	Flow rate	
cubic foot per day (ft^3/d)	0.02832	cubic meter per day (m^3/d)
foot per day (ft/d)	0.3048	meter per day (m/d)
inch per year (in/yr)	25.4	millimeter per year (mm/yr)
	Transmissivity*	
foot squared per day (ft^2/d)	0.09290	meter squared per day (m^2/d)

Vertical leakance is measured in d^{-1}.

*Transmissivity: The standard unit for transmissivity is cubic foot per day per square foot times foot of aquifer thickness [(ft^3/d)/ft^2]ft. In this report, the mathematically reduced form, foot squared per day (ft^2/d), is used for convenience.

Abbreviations

GWM Groundwater Management (Process)

LP linear programming

MNW multi-node well

RMS Response Matrix Solution (Package)

SLP sequential linear programming

WEL Well (Package)

Use of Multi-Node Wells in the Groundwater-Management Process of MODFLOW–2005 (GWM–2005)

By David P. Ahlfeld[1] and Paul M. Barlow[2]

Abstract

Many groundwater wells are open to multiple aquifers or to multiple intervals within a single aquifer. These types of wells can be represented in numerical simulations of groundwater flow by use of the Multi-Node Well (MNW) Packages developed for the U.S. Geological Survey's MODFLOW model. However, previous versions of the Groundwater-Management (GWM) Process for MODFLOW did not allow the use of multi-node wells in groundwater-management formulations. This report describes modifications to the MODFLOW–2005 version of the GWM Process (GWM–2005) to provide for such use with the MNW2 Package. Multi-node wells can be incorporated into a management formulation as flow-rate decision variables for which optimal withdrawal or injection rates will be determined as part of the GWM–2005 solution process. In addition, the heads within multi-node wells can be used as head-type state variables, and, in that capacity, be included in the objective function or constraint set of a management formulation. Simple head bounds also can be defined to constrain water levels at multi-node wells. The report provides instructions for including multi-node wells in the GWM–2005 data-input files and a sample problem that demonstrates use of multi-node wells in a typical groundwater-management problem.

Introduction

The Groundwater-Management (GWM) Process developed for the modular three-dimensional groundwater model MODFLOW provides a set of optimization-modeling techniques that can be used to solve several types of groundwater-management problems, such as maximizing groundwater withdrawals from an aquifer or minimizing streamflow depletions caused by pumping (Ahlfeld and others, 2005, 2009, and 2011). Each management (or optimization) problem consists of a set of decision variables, an objective function that is maximized or minimized, and a set of constraints that limit the optimal value of the objective function. The decision variables of a management formulation are the quantifiable controls (or decisions) that are determined as part of the solution to the problem. The primary type of decision variable—a flow-rate decision variable—that is supported by GWM is a withdrawal (discharge) or injection (recharge) flow rate at a managed well site. In previous versions of GWM, these withdrawal or injection flow rates were simulated by use of MODFLOW's standard WEL Package (Harbaugh, 2005). In the WEL Package, each simulated well is assumed to be connected to a single cell (or node) of the model grid that represents the groundwater-flow system, and the water level in the well is assumed to be identical to the water level (or head) at the connected node.

The multi-node well packages of MODFLOW (MNW1, Halford and Hanson, 2002; MNW2, Konikow and others, 2009), however, allow simulation of wells that extend beyond a single model node. This capability provides a means to simulate wells that penetrate more than one model layer or more than one aquifer, or to represent wells that are slanted or horizontal. In addition, the single water level that is calculated for a multi-node well does not need to equal (and most often will not equal) the water levels at each of the connected nodes. Flow rates to and from each connected node will vary in magnitude (and perhaps in direction) along the length of the open or screened intervals of the well on the basis of the relative heads and hydraulic conductances between each of the connected nodes and the water level in the well (Konikow and others, 2009). The net flow rate into or out of the multi-node well from all connected nodes represents the net flow of the well at the wellhead. This flow rate can be negative (representing a withdrawal well), positive (representing an injection well), or zero (representing a nonpumping well or long-screened monitoring well).

[1]University of Massachusetts Amherst.

[2]U.S. Geological Survey.

In prior versions of GWM, it was possible to include MNW-type wells as unmanaged wells—that is, as background pumping stresses that are part of the groundwater-flow process simulation but not part of the management formulation. This report describes modifications to the GWM Process to extend the use of MNW-type wells as part of the management formulation in addition to their use as unmanaged wells. MNW-type wells can be incorporated into a management formulation in two ways: (1) as flow-rate decision variables, in which the flow rate at each managed MNW-type well is determined as part of the optimization process; and (2) as sites for management of water levels, at which the water level at an MNW-type managed well is constrained by a head-type constraint or managed by use of head-type state variables. Both management options can be simulated simultaneously at an MNW-type well; in other words, it is possible to manage water levels at an MNW-type flow-rate decision variable. The report describes the use of multi-node wells in GWM management formulations, a sample problem that demonstrates the new capabilities, and data-input instructions for including multi-node wells in a GWM problem.

The modifications that are described here are applicable only to the MODFLOW–2005 (Harbaugh, 2005) version of the GWM Process, which is referred to as GWM–2005; all further references to GWM in this report are to the 2005 version of the software. Also, because the MNW2 Package is an update to the original MNW1 Package, only the MNW2 Package (described in Konikow and others, 2009) has been included in the GWM Process; therefore, all further references to MNW are to the MNW2 Package or, more generally, to multi-node wells. It is assumed that the reader is familiar with MODFLOW–2005, the MNW2 Package, and GWM–2005. Only enough background information is provided in the report to understand how the MNW2 functionality has been implemented in GWM–2005. Much of the report is focused on concepts and information that are needed by users of GWM to understand and correctly prepare the data-input files for a GWM simulation. Detailed line-by-line instructions for preparing the required data-input files are given in the appendix at the back of this report.

In the remainder of the report, wells simulated by MODFLOW's WEL Package are referred to as "WEL-type" wells, and those simulated by the MNW2 Package as "MNW-type" wells. Also, throughout the report, names for all file types are shown in bold uppercase text, such as MODFLOW's **NAME** file.

Multi-Node Wells in Management Formulations

As noted previously, MNW wells can be used in GWM as either unmanaged or managed wells. If they are used as unmanaged wells (that is, simulated as part of the background groundwater-flow system), they are specified in the data-input file for the MNW2 Package. If they are used as part of the management formulation, then they must be defined as flow-rate decision variables in the GWM decision-variables (**DECVAR**) file; minimum and maximum pumping (or injection) rates at each MNW well must be defined in the GWM decision-variable constraints (**VARCON**) file.

Multi-Node Wells Defined as Flow-Rate Decision Variables

This section describes the data-input requirements necessary to include MNW-type wells as flow-rate decision variables in a GWM formulation. Important issues related to the use of MNW-type wells in a GWM management problem are the definition and characteristics of the simulated flow rate at an MNW-type well. For wells simulated by use of MODFLOW's WEL Package, the user explicitly specifies a flow rate (Q) that represents the total flow rate to or from a single grid cell. In contrast, for MNW-type wells, the user specifies the maximum desired (Q_{des}) net flow rate for each MNW-type well. This flow rate is referred to as a "desired" flow rate because the actual flow rate at the wellhead can be constrained by the inability of the aquifer to supply the flow needed to meet the "desired" flow rate. The desired flow rate also can be constrained by user-specified maximum and minimum water levels at the well and by possible pump-capacity constraints, as described in Konikow and others (2009) and later in this section of the report. The actual flow rate that is determined by the MNW2 Package solution process is then distributed among the multiple connected nodes that make up the well. The MNW2 Package accomplishes this distribution by first defining a new head—the head in the multi-node well. A flow rate to or from each node that contains the well can then be computed based on the difference between the head in the cell and the head in the well and on the hydraulic conductance between the cell and well. The MNW2 Package uses an iterative procedure to find a single water level in the well that will produce flows between each node and the well whose sum equals the total flow rate at the wellhead.

When an MNW-type well is defined as a managed well, GWM will determine an optimal flow rate for the well as part of the solution process. Therefore, the user does not specify a desired flow rate for managed MNW-type wells. Instead, GWM searches for a flow rate at the managed MNW-type well that optimizes the management formulation. At the same time, GWM ensures that the optimal flow rate at the MNW-type well will still satisfy the conditions required by the MNW2 Package. During the GWM solution process, the MNW2 Package is provided with tentative desired flow rates (Q_{des}) at each managed MNW well by the GWM algorithm; the MNW2 Package then calculates a single water level at the well and balances flows among the connected nodes that make up the well.

The difference in the way that MNW-type and WEL-type flow-rate decision variables are handled in GWM affects the Volumetric Budget output section of the MODFLOW **LIST** file. GWM–2005 writes a separate line for volumes and rates associated with WEL-type decision variables. This line is labeled "Managed Wells" in the Volumetric Budget section. Volumes and rates for managed and unmanaged MNW wells are written to a single line by the MNW2 Package in a line labeled "MNW2."

Defining the Number and Locations of Managed Multi-Node Wells

The combined total number of MNW-type and WEL-type flow-rate decision variables defined for a GWM run is specified by use of the variable NFVAR in item 2 of the **DECVAR** file (see the appendix at the back of this report for a complete listing of the input instructions for the use of MNW-type wells in a GWM formulation). Information about each flow-rate variable is then defined by use of items 3a through 3e. GWM determines the types of wells that are defined for each flow-rate variable by the value specified for the variable NC in item 3a: if NC is specified as a positive integer, then the variable consists of one or more WEL-type wells, each of which consists of a single model cell; if NC is specified as a negative integer, then the variable consists of one or more MNW-type wells, each of which can consist of one or more model cells. A flow-rate decision variable cannot consist of a mix of WEL-type and MNW-type wells; the multiple wells defined by the decision variable must be either all WEL-type or all MNW-type wells.

Previous versions of GWM allowed the user to define WEL-type flow-rate decision variables to extend over multiple cells (that is, the variable NC specified to be greater than or equal to 2). This specification can be useful, for example, for situations in which a single optimal pumping rate is sought for a group of wells in an irrigation district or water-supply system. This capability is retained with the updated software. When NC is specified to be greater than or equal to 2, the user specifies the fraction of the total flow rate for the decision variable that is distributed to each of the NC model cells (Appendix, Decision Variables (**DECVAR**) file, item 3b-1). The fraction of flow distributed to a cell is specified by use of the variable RATIO, and the sum of the RATIO values specified for all cells by the decision variable must equal 1.0.

With the use of MNW-type wells, the variable NC takes on an extended meaning. If the flow-rate decision variable consists of a single MNW-type well, then NC is specified as -1, and the number of model cells associated with the multi-node well is defined by the use of the variable NNODES in item 3b-2. If the flow-rate decision variable consists of more than one MNW-type well (for example, a group of wells within an irrigation district), then NC is specified to equal the negative of the total number of these wells. Item 3b-3 is then used to define the fraction of the total flow rate that is distributed to each multi-node well in the decision variable (the variable RATIO) and the number of model cells associated with each multi-node well in the decision variable (the variable NNODES). GWM then determines a single optimal withdrawal (or injection) rate for the group of wells defined for the decision variable. A fraction of this optimal flow rate is distributed to each MNW-type well in the decision variable according to the values specified for the variable RATIO for each well. Each MNW-type well, in turn, distributes its assigned flow rate to its multiple nodes according to the solution algorithm in the MNW2 Package to produce a single head value in each MNW-type well.

The MNW2 Package provides two options for specifying the locations of model grid cells associated with each multi-node well (Konikow and others, 2009). The user does not need to apply the same option to all wells. In the first approach, the user specifies the layer, row, and column numbers of each node associated with a multi-node well; in the second approach, the user specifies the elevations of the tops and bottoms of the open intervals (or well screens) of a multi-node well in addition to the row and column numbers of the well. Option two is applicable only to vertical wells. The variable NNODES in record 3b is used to indicate which option is being used. If NNODES is specified to be greater than zero, then the first option is selected, and the user completes input item 3e-1. If NNODES is specified to be less than zero, then the second option is selected, and the user completes input item 3e-2. Additional details for specifying the locations of multi-node wells are provided in the input instructions for the **DECVAR** file in the appendix at the back of this report and in Konikow and others (2009).

Head-Loss Terms and Partial Penetration of Well Screens

As noted previously, the MNW2 Package calculates the composite head (or water level) for each multi-node well in addition to heads at each node in which the well is located. To account for possible differences in water levels between the well and each node, the following well-loss equation is used by the package to relate the head calculated at each node of the MODFLOW grid associated with the well (h_n) to the composite head (or water level) in the well (h_{well}):

$$h_{well} = h_n + [AQ_n + BQ_n + CQ_n^P], \quad (1)$$

where

Q_n is the flow between the *n*th node and the well,
A is a linear aquifer-loss coefficient,
B is a linear well-loss coefficient,
C is a nonlinear well-loss coefficient, and
P is the power (exponent) of the nonlinear discharge component of well loss.

In accordance with standard MODFLOW practice, Q_n is taken to be negative for a pumping (discharge) well. The first term in the brackets on the right side of equation 1 (AQ_n) accounts for head losses in the aquifer that result from the well having a radius that is smaller than the horizontal dimensions of the cell in which the well is located (that is, cell-to-well head losses). The second term (BQ_n) accounts for head losses caused by skin effects adjacent to and within the borehole and well screen, and the third term (CQ_n^P) accounts for nonlinear head losses caused by turbulent flow near the well. Additional information for each of the three coefficients and nonlinear exponent in equation 1 is provided in Konikow and others (2009).

The user specifies which types of well losses should be simulated at a particular multi-node well by use of the variable LOSSTYPE in record 3c of the **DECVAR** file. There are five options:

a. If LOSSTYPE is specified as NONE, there are no well-loss corrections, and the head in the well is assumed to equal the head at the node ($h_{well} = h_n$); this option is valid only for a single-node well (that is, NNODES specified as 1).

b. If LOSSTYPE is specified as THIEM, only the cell-to-well head-loss correction is simulated ($h_{well} = h_n + AQ_n$); in this case, the user specifies the radius of the well (Rw) in items 3d or 3e. This option is referred to as THIEM because it is based on the Thiem steady-state flow correction, as described in Konikow and others (2009).

c. If LOSSTYPE is specified as SKIN, then both the cell-to-well head-loss correction and the effects of formation damage or skin effects are simulated ($h_{well} = h_n + AQ_n + BQ_n$); in this case, the user specifies the radius of the well (Rw) and the radius (Rskin) and hydraulic conductivity (Kskin) of the skin in items 3d or 3e.

d. If LOSSTYPE is specified as GENERAL, then all three head-loss corrections are simulated, and the user specifies the variables Rw, B (the linear well-loss coefficient), C (the nonlinear well-loss coefficient), and P (the nonlinear exponent) in items 3d or 3e.

e. Finally, if LOSSTYPE is specified as SPECIFYcwc, then the user specifies an effective cell-to-well conductance value (CWC) that accounts for all well-loss corrections in items 3d or 3e; the variable CWC is defined by equation 15 in Konikow and others (2009).

The MNW2 Package allows the well-loss variables associated with equation 1 (that is, Rw, Rskin, Kskin, B, C, P, and CWC) to be constant values for all nodes associated with a particular multi-node well or to vary from one node (or interval) to the next. For the former condition, the user specifies the well-loss variables in item 3d and for the latter condition in item 3e.

It is sometimes the case that an open or screened interval of a well does not penetrate the full thickness of the aquifer or does not span the full thickness of the simulated model layer that contains the open interval. Such partial penetration can create vertical-flow components that cause additional drawdowns at the well that are not accounted for by the well-loss terms described in equation 1. The MNW2 Package provides an option to correct for partial-penetration effects on a model-layer basis, and this option is retained in the GWM Process. Konikow and others (2009) provide a detailed description of how partial-penetration effects are simulated by the package, the assumptions that underlie implementation of the partial-penetration correction in the package, and the conditions under which the partial-penetration correction may cause model-convergence problems. The option is implemented for an MNW-type well by specifying a value greater than zero for the variable PPFLAG in item 3c and the fraction of partial penetration of the well within a model cell by use of the variable PP in item 3e. Users should be cautioned, however, that the partial-penetration option in GWM has not been extensively tested; users should closely evaluate whether or not the added simulation capability offered by the partial-penetration correction outweighs potential convergence problems.

Constraining Pumping Rates at Managed Multi-Node Wells

The MNW2 Package has several options to control the pumping rate at an MNW-type well. These include direct control by specifying the minimum and maximum pumping rates for a well and indirect control by specifying the minimum or maximum water levels within a well. Each of the options is implemented in the MNW2 input file by first specifying any nonzero value to the variable Qlimit and, subsequently, appropriate values for variables Qfrcmn, Qfrcmx, and Hlim. The variable Qfrcmn is used by the MNW2 Package to represent the lower limit of the fixed range of withdrawal capacity for a well; discharge at the well is reduced to zero if the computed net discharge at the well falls below Qfrcmn. Pumping at the well can be restored if the potential pumping rate exceeds the value specified for Qfrcmx (Qfrcmx must be greater than Qfrcmn). The variable Hlim is used by the MNW2 Package to represent the minimum water level at a pumped well (or maximum water level for a recharge well). When the water level reaches or falls below Hlim, discharge from (or recharge to) the well is reduced.

These options are retained in GWM for MNW-type wells simulated as unmanaged wells but are not available when MNW-type wells are defined as managed wells. For the latter case, GWM forces Qlimit to take a value of zero, which indicates to the MNW2 Package that none of the pumping constraints will be imposed. Instead, constraints on pumping rates at MNW-type wells that have been defined as flow-rate decision variables are imposed as part of the management formulation. Minimum and maximum pumping rates at each MNW-type managed well can be specified by use of variables FVMIN and FVMAX in the **VARCON** file, as they are for WEL-type managed wells. The capabilities to stop and restart pumping at an MNW-type well represented by Qfrcmn and Qfrcmx cannot be simulated by GWM; however, the user can define a nonzero minimum pumping rate at an MNW-type well either by associating a binary variable with the well or by the use of summation constraints, as described by Ahlfeld and others (2005, p. 31).

Unsupported Options of the MNW2 Package for Managed Multi-Node Wells

In addition to the options provided by the variable Qlimit to control the pumping rate at an MNW-type well, two other variables in the MNW2 Package that describe the behavior of the pump are not available for managed MNW-type wells (although the options are available for unmanaged MNW-type wells). The first is the variable PUMPLOC, which is used to define the actual location of the pump intake within a well. Because this variable affects only the routing of flow and solutes within a borehole and not the flow rates to or from a well, PUMPLOC is not necessary for a GWM simulation and is therefore set to the default value of zero by GWM. The second variable is PUMPCAP, which specifies whether the discharge of a pumping well will be adjusted for changes in pumping lift (or total dynamic head at the well) with time. GWM forces PUMPCAP to equal zero for managed MNW-type wells to prevent potential nonlinear effects.

Management of Water Levels at Multi-Node Wells

The MNW2 Package calculates the water level (or head) in each multi-node well at each time step. MNW-type wells that are unmanaged (that is, specified in an MNW2 Package input file) retain all of the functionality provided with the MNW2 Package, such as controlling the water level at a well by use of the variable Hlim. However, if the MNW-type well is not managed, the water level in the well cannot be part of the management formulation specified in GWM.

Water levels at MNW-type wells that have been defined as managed wells in the GWM decision-variables input file can be managed in two ways. First, maximum and minimum head-bound constraints can be imposed on the water level at a managed MNW-type well by specification of appropriate input values in the head-constraints (**HEDCON**) input file. Second, water levels at managed MNW-type wells can be defined as head-type state variables, and therefore can be used in the objective function or summation constraints of a GWM formulation. Instructions for including water levels at managed MNW-type wells in the GWM input files are provided in the appendix at the back of this report.

Sample Problem: MNW–Supply

The purpose of this sample groundwater-management problem is to demonstrate use of GWM with managed and unmanaged multi-node wells. The hypothetical aquifer system that is the basis for this problem is similar to the system described by Halford and Hanson (2002).

The simulated system consists of two aquifers separated by a confining unit. The upper aquifer (layer 1 of the model) is unconfined with a hydraulic conductivity of 60 feet per day (ft/d) and a uniform base of 50 feet (ft) above an arbitrary datum of 0 ft. The lower aquifer (layer 2 of the model) is confined with a hydraulic conductivity of 300 ft/d and a uniform base of 0 ft. The confining unit is not explicitly simulated but instead is represented in the model by a vertical leakance of 2.0×10^{-4} $days^{-1}$ (d^{-1}) between the two model layers. The top layer of the model is assigned a specific yield of 0.05 and the bottom layer a storage coefficient of 1.0×10^{-4}.

The model grid consists of 21 rows and 14 columns, and grid cells have a uniform length of 2,500 ft on each side (fig. 1). A total simulation period of 10,970 days (d) is simulated with five stress periods. The first two stress periods, which are both steady state, are each 5,000 d in length. The third through fifth stress periods are transient with lengths of 60, 180, and 730 d, respectively.

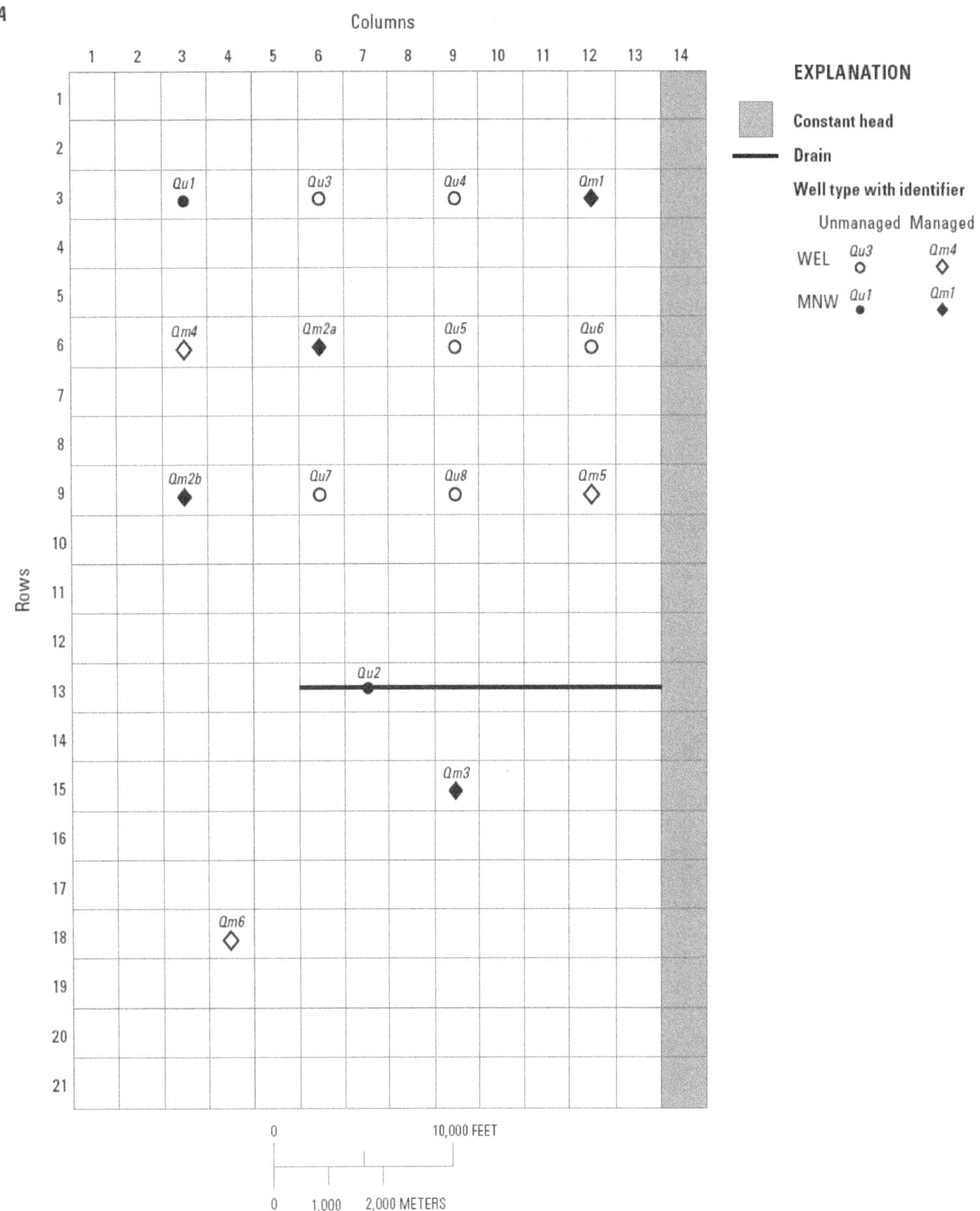

Figure 1. The model grid for the MNW–SUPPLY sample problem showing grid-cell locations of *A*, managed and unmanaged wells and *B*, specified-head constraints.

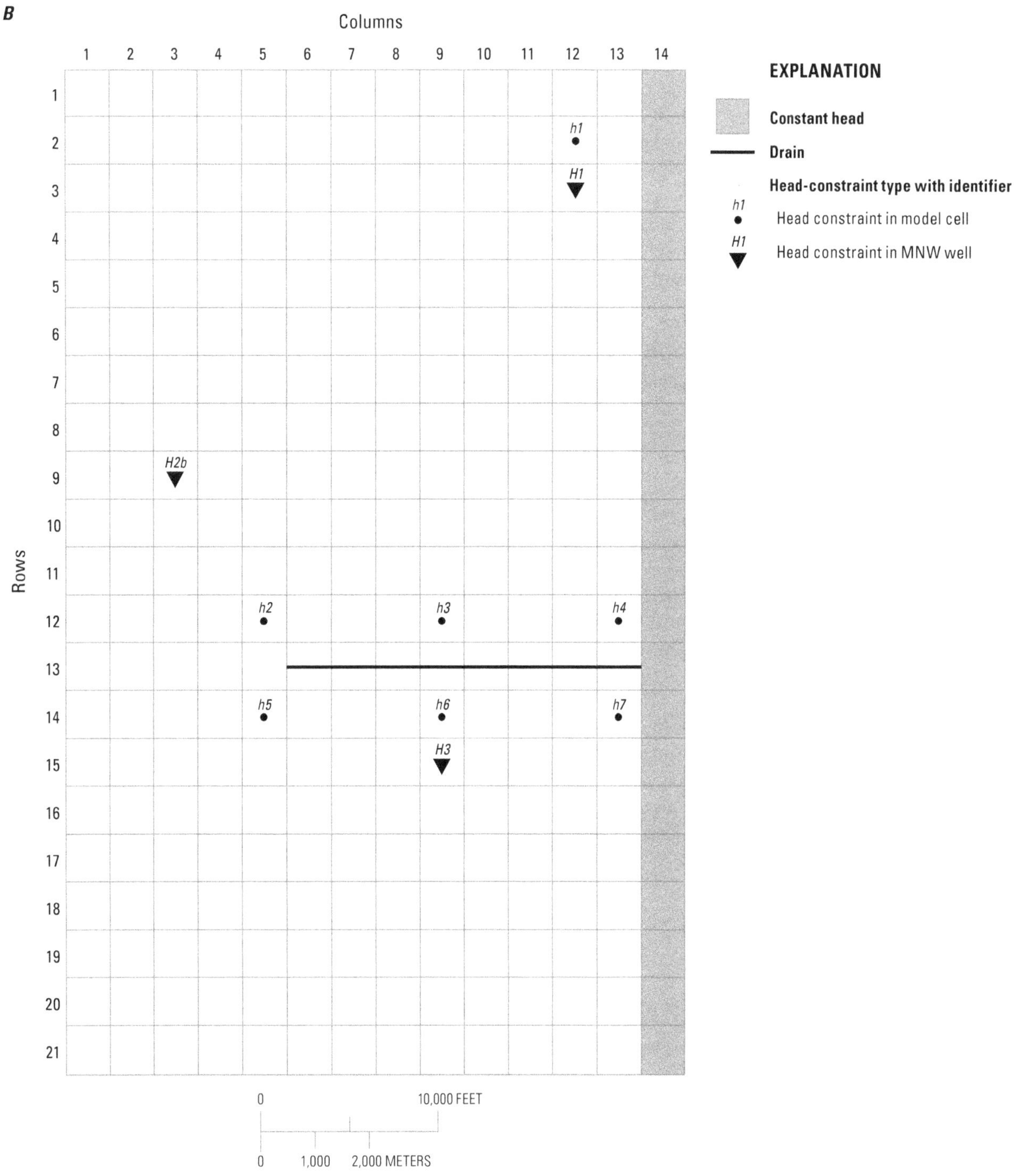

Figure 1. The model grid for the MNW–SUPPLY sample problem showing grid-cell locations of *A,* managed and unmanaged wells and *B,* specified-head constraints.—Continued

The groundwater system is bounded on the right by a stream that is simulated by use of constant heads in the top layer of the model. The specified heads range from a maximum of 139 ft at the top right boundary of the model to a minimum of 119 ft on the bottom right of the model; these heads are constant throughout the simulation. A drain is also specified in the top layer of the model. Elevations of the drain cells range from a maximum of 131 ft on the left end of the drain to a minimum of 128 ft on the right end. Recharge to the upper aquifer is simulated at rates of 7 inches per year (in/yr; 0.0016 ft/d) for the first two stress periods, 2 in/yr (0.000457 ft/d) for the third stress period, 0 in/yr for the fourth stress period, and 12 in/yr (0.0028 ft/d) for the fifth stress period. The general direction of groundwater flow in the absence of pumping is from the left side of the model domain toward the discharge boundaries at the stream and drain.

The model domain includes a total of 15 well sites (fig. 1A), and the wells are pumped from both the confined and unconfined model layers. Pumping rates at 8 of the wells (identified as wells $Qu1$ through $Qu8$ on fig. 1A) are specified in the model (as unmanaged pumping rates); pumping rates at the remaining wells ($Qm1$ through $Qm6$ on fig. 1A) are to be determined as part of the management problem. The wells consist of a mix of 9 WEL-type and 6 MNW-type wells (fig. 1A).

The MODFLOW model consists of the following input files: **NAME**, Basic (**BAS**), Discretization (**DIS**), Block-Centered Flow (**BCF**), Time-Variant Specified Head (**CHD**), Drain (**DRN**), Well (**WEL**), Recharge (**RCH**), Multi-Node Well (**MNW2**), Output Control (**OC**), and Preconditioned Conjugate-Gradient (**PCG**).

The objective of the management problem is to maximize the volume of water pumped at the managed wells during the second through fifth stress periods. The objective function is stated mathematically as

$$Maximize\ T_{Qm1}Qm1 + T_{Qm2}Qm2 + T_{Qm3}Qm3 + T_{Qm4}Qm4 + T_{Qm5}Qm5 + T_{Qm6}Qm6\,. \qquad (2)$$

Variables shown as T_{Qmn} in equation 2, where subscript n is the well number, represent the total duration of pumping at each of the decision variables. Although equation 2 includes a total of six decision variables, pumping is to be managed at seven well sites because decision variable $Qm2$ is defined to consist of two MNW-type wells (wells $Qm2a$ and $Qm2b$ on fig. 1A). Each of these two wells is assigned 50 percent of the optimal pumping rate for $Qm2$ (that is, input variable RATIO in the **DECVAR** input file is specified as 0.5 for each well).

The managed wells consist of both WEL-type and MNW-type wells (table 1). As shown in the table, the MNW wells are assigned different types of well losses (SKIN, THIEM, and SPECIFYcwc) and both options for specifying the vertical location of each well (by cell or by vertical interval). Different characteristics are given to the wells to illustrate the various input options for MNW-type wells. Table 1 also lists the stress periods during which each of the decision variables is active and the upper bound on pumping rate for each of the decision variables. Note that the decision-variable pumping rates are constant during the stress periods listed in table 1. Additional details on the well-loss characteristics and screened intervals of each well are shown in the **DECVAR** input file for the problem provided in figure 2.

Table 1. Characteristics of managed wells in the MNW–SUPPLY sample problem.

[--, not applicable; upper bounds on pumping rates are in cubic feet per day]

Decision variable		Well type	Method of vertical discretization	Layers spanned by well	Horizonal grid-cell location		Stress periods associated with decision variable	Loss type	Upper bound on pumping rate
					Row	Column			
Qm1		MNW	Cell	1–2	3	12	2–3	SKIN	20,000
Qm2[1]							2–5		400,000
	Qm2a	MNW	Interval	1	6	6	--	THIEM	--
	Qm2b	MNW	Interval	1	9	3	--	THIEM	--
Qm3		MNW	Interval	2	15	9	2–5	SPECIFYcwc	600,000
Qm4		WEL	Cell	1	6	3	3–5	--	400,000
Qm5		WEL	Cell	1	9	12	2–5	--	20,000
Qm6		WEL	Cell	1	18	4	2–5	--	400,000

[1]Decision variable Qm2 comprises wells Qm2a and Qm2b.

```
#MNW-SUPPLY Sample Problem, DECVAR file
#January 2012
1 30                            Item 1
6 0 0                           Item 2
Qm1  -1  1  1  1  W  Y  2:3     DV 1: Item 3a: NC = -1
MNWQm1  2                    Item 3b-2: WELLID, NNODES
SKIN  0                      Item 3c: LOSSTYPE, PPFLAG
-1  -1  -1                   Item 3d: -Rw, -Rskin, -Kskin
1 3 12 0.500 1.00 24.563     Item 3e-1, node 1: L,R,C, Rw,Rskin,Kskin
2 3 12 0.500 1.00 122.815    Item 3e-1, node 2: L,R,C, Rw,Rskin,Kskin
Qm2  -2  1  1  1  W  Y  2:3:4:5  DV 2: Item 3a: NC = -2
0.5 MNWQm2a  -1             Item 3b-3, first well: RATIO,WELLID,NNODES
THIEM  0                    Item 3c, first well: LOSSTYPE, PPFLAG
 0.5                         Item 3d, first well: Rw
450 50 6 6                  Item 3e-2, first well: Ztop,Zbotm,Row,Col
0.5 MNWQm2b  -1             Item 3b-3, second well: RATIO,WELLID,NNODES
THIEM  0                    Item 3c, second well: LOSSTYPE, PPFLAG
 0.5                         Item 3d, second well: Rw
250 100 9 3                 Item 3e-2, second well: Ztop,Zbotm,Row,Col
Qm3  -1  1  1  1  W  Y  2:3:4:5  DV 3: Item 3a: NC = -1
MNWQm3  -2                      Item 3b-2: WELLID, NNODES
SPECIFYcwc  0                   Item 3c: LOSSTYPE, PPFLAG
5000                             Item 3d: CWC
50 30 15 9                      Item 3e-2, node 1: Ztop,Zbotm,Row,Col
25 10 15 9                      Item 3e-2, node 2: Ztop,Zbotm,Row,Col
Qm4   1    1    6    3    W    Y    3:5        DV 4: Item 3a: NC=1
Qm5   1    1    9    12   W    Y    2:3:4:5    DV 5: Item 3a: NC=1
Qm6   1    1    18   4    W    Y    2:3:4:5    DV 6: Item 3a: NC=1
```

Figure 2. Decision-variables (**DECVAR**) file for the MNW–SUPPLY sample problem.

The value of the objective function is limited by several constraints specified in the GWM input files. First, lower and upper bounds are specified on the pumping rate of each decision variable in the GWM **VARCON** file (fig. 3); the lower bound on the pumping rate at each well is zero (shown by the column for variable FVMIN in fig. 3), and the upper bounds are shown in table 1 and by the column for variable FVMAX in figure 3. Note that although a maximum pumping rate is specified by FVMAX for each MNW-type well, these rates are not the same as the desired pumping rates (Q_{des}) specified for an unmanaged MNW-type well. For these managed MNW-type wells, the maximum pumping rate at each well will be limited by the values of FVMAX but determined by the optimization process.

The remaining constraints consist of a series of head-type constraints specified at seven model cells and three MNW well locations. The locations of these head-type constraints are shown in figure 1B. The first constraints are simple minimum head-bound constraints specified by use of a **HEDCON** file. These constraints are written mathematically as

$$h_{cell} \geq h^*_{cell} \tag{3a}$$

for heads specified at model cells and as

$$h_{MNWwell} \geq h^*_{MNWwell} \tag{3b}$$

```
#MNW-SUPPLY Sample Problem, VARCON file
#January 2012
 1                                 #1-IPRN
 Qm1        0     2d4   0           #2-FVNAME FVMIN FVMAX FVREF
 Qm2        0     4d5   0
 Qm3        0     6d5   0
 Qm4        0     4d5   0
 Qm5        0     2d4   0
 Qm6        0     4d5   0
```

Figure 3. Decision-variable constraints (**VARCON**) file for the MNW–SUPPLY sample problem.

for heads specified at MNW-type wells. Variables h_{cell} and $h_{MNWwell}$ represent heads calculated by GWM at model cells and MNW wells, respectively. Variables h^{*}_{cell} and $h^{*}_{MNWwell}$ represent minimum heads specified in the **HEDCON** file at model cells and MNW wells, respectively. A total of 36 head constraints are specified at model cells (12 each in stress periods 3, 4, and 5) and 3 head constraints at MNW wells (1 in stress period 2 and 2 in stress period 5) (fig. 4A).

The second type of head constraint is a head-difference constraint in the third period of the simulation that is specified between the head in managed well $Qm1$ (defined as $H1$ and shown on fig. 1B) and the head in one of the adjoining cells (defined as $h1$ and shown on fig. 1B). The constraint requires that the head calculated at the managed well at the end of stress period three be no greater than 5 ft lower than the head calculated at the adjoining cell:

$$h_1 - H_1 \leq 5.0 \,. \tag{4}$$

This head-difference constraint was specified by defining each of the heads as state variables in the state-variables (**STAVAR**) file (fig. 4B) and then defining a summation constraint with the state variables in the summation-constraints (**SUMCON**) file (fig. 4C).

Because of the presence of the water table in layer 1 of the model and the head-dependent boundary condition at the simulated drains, there is a potential nonlinear relation between pumping rates at the wells and head changes in the groundwater system. For this reason, the nonlinear sequential linear programming (SLP) solution process of GWM is used to solve the management formulation (Ahlfeld and others, 2005). The problem is solved in just three iterations of the SLP process for closure criteria of 1.0×10^{-5} cubic foot per day (ft^3/d) specified for the maximum change in pumping rate at each decision variable from one iteration to the next and 1.0×10^{-4} cubic foot (ft^3) specified for the maximum change in the value of the objective function from one iteration to the next.

The value of the objective function at the optimal solution is 2.72×10^8 ft^3 (fig. 5) with pumping from decision variables $Qm1$ (20,000.0 ft^3/d), $Qm2$ (5,681.7 ft^3/d), $Qm3$ (2,971.3 ft^3/d), and $Qm5$ (20,000.0 ft^3/d). Decision variables $Qm1$ and $Qm5$ are pumped at their maximum rates because they are close to the constant-head boundary at the simulated stream and therefore have relatively small effects on the head constraints. In addition to the upper bounds on pumping at these two wells are two simple head-bound binding constraints to the management problem at site h_2 (in layer 2 for stress period 5) and site h_7 (layer 2 for stress period 5) (fig. 5). However, the head-difference constraint specified between the head in managed well $Qm1$ (shown as H1_SP3 in fig. 5) and the head in one of the adjoining cells (shown as h1_SP3 in the figure) is not binding. The difference in heads between the two locations is 1.9662 ft (150.2377 ft at h1_SP3 and 148.2715 ft at H1_SP3), which is less than the prescribed maximum of 5.0 ft defined in equation 4.

Details about the pumping rates at unmanaged and managed wells are found in the MODFLOW **LIST** file for the sample problem, a portion of which is shown for stress period 2 in figure 6. The first part of the figure shows the specified pumping (stress) rates at unmanaged wells $Qu3$ through $Qu8$ (fig. 6A), which are WEL-type wells whose pumping rates are read from the MODFLOW WEL file. The second part of the figure shows the managed flow rates at wells $Qm5$ and $Qm6$, which are WEL-type wells whose pumping (stress) rates are determined by GWM (fig. 6B). The third part of the figure shows summary information for the unmanaged ($Qu1$ and $Qu2$) and managed ($Qm1$, $Qm2a$, $Qm2b$, and $Qm3$) MNW-type wells (fig. 6C).

(A) The head-constraints file

```
#MNW-SUPPLY Sample Problem, HEDCON file
#January 2012
1      #1-IPRN
39     0      0      0      #2-NHB NDD    NDF    NGD
h2_L1_SP3    1      12     5      ge     159    3
h3_L1_SP3    1      12     9      ge     145    3
h4_L1_SP3    1      12     13     ge     134    3
h5_L1_SP3    1      14     5      ge     152    3
h6_L1_SP3    1      14     9      ge     143    3
h7_L1_SP3    1      14     13     ge     133    3

24 lines deleted here (see input file 'mnwsupply.hedcon' for the complete list of
constraints)

h2_L2_SP5    2      12     5      ge     168    5
h3_L2_SP5    2      12     9      ge     157    5
h4_L2_SP5    2      12     13     ge     148    5
h5_L2_SP5    2      14     5      ge     168    5
h6_L2_SP5    2      14     9      ge     157    5
h7_L2_SP5    2      14     13     ge     148    5
H2b_SP2      0  MNWQm2b ge   40   2
H2b_SP5      0  MNWQm2b ge   76   5
H3_SP5       0  MNWQm3  ge   65   5
```

(B) The state-variables file

```
#MNW-SUPPLY Sample Problem, STAVAR file
#January 2012
 1                            #1-IPRN
 2  0  0  0                   #2-NHVAR NRVAR NSVAR NDVAR
 H1_SP3   0 MNWQm1   3        # SVNAME MFLG WELLID (since MFLG=0) SVSP
 h1_SP3   2   2  12  3        # SVNAME LAY ROW COL SVSP
```

(C) The summation-constraints file

```
#MNW-SUPPLY Sample Problem, SUMCON file
#January 2012
 1                   #1-IPRN
 1                   #2-SMCNUM
 MNW_1_HD   2 le 5.0    #3a-SMCNAME NTERMS TYPE RHS
  h1_SP3  1.0        #3b-GVNAME GVCOEFF
  H1_SP3 -1.0
```

Figure 4. Listings of the *A*, head-constraint (**HEDCON**), *B*, state-variables (**STAVAR**), and *C*, summation-constraint (**SUMCON**) files for the MNW–SUPPLY sample problem.

```
-----------------------------------------------------------------------
              Groundwater Management Solution
-----------------------------------------------------------------------

       OPTIMAL SOLUTION FOUND

       OPTIMAL RATES FOR EACH FLOW VARIABLE
       ---------------------------------------
Variable           Withdrawal          Injection           Contribution
Name               Rate                Rate                To Objective
----------         --------------      ------------        ------------
 Qm1               2.000000E+04                            1.012000E+08
 Qm2               5.681737E+03                            3.392054E+07
 Qm3               2.971347E+03                            1.773924E+07
 Qm4               0.000000E+00                            0.000000E+00
 Qm5               2.000000E+04                            1.194020E+08
 Qm6               0.000000E+00                            0.000000E+00
                   ------------        ------------        ------------
TOTALS             4.865308E+04        0.000000E+00        2.722618E+08

       OPTIMAL VALUES FOR EACH STATE VARIABLE
       ---------------------------------------
Variable                                                   Contribution
Name               Value                                   To Objective
----------         ------------                            ------------
 H1_SP3             1.482715E+02                            0.000000E+00
 h1_SP3             1.502377E+02                            0.000000E+00
                   ------------                            ------------
TOTALS              2.985092E+02                            0.000000E+00


       OBJECTIVE FUNCTION VALUE                            2.722618E+08

       BINDING CONSTRAINTS
Constraint Type        Name      Status       Shadow Price
---------------        ----      ------       ------------
Head Bound             h2_L2_SP5Binding        -2.5338E+08
Head Bound             h7_L2_SP5Binding        -2.4723E+08
Maximum Flow Rate      Qm1       Binding      Not Available
Maximum Flow Rate      Qm5       Binding      Not Available
```

Figure 5. Selected output from the GWM **OUT** file for the optimal solution of the MNW–SUPPLY sample problem.

Flow rates calculated for each MNW-type well include both inflow (*Qin*) and outflow (*Qout*) components, as well as the net flow to and from each MNW well shown in the column identified as *Qnet*. Note that there is a small outflow rate from well *Qu*1 to the aquifer (575.2 ft^3/d). Also note in this section that the net discharge rates at wells *Qm*2*a* and *Qm*2*b* (2,840.87 ft^3/d) are half the total rate from decision variable *Qm*2 (fig. 5; 5.681710E+03 ft^3/d), which is consistent with the values of RATIO (0.5) specified for each of the wells in the decision-variables input file for the management problem. The fourth part of the figure (fig. 6D) shows the overall inflows and outflows to and from the unmanaged and managed wells as part of the volumetric budget for the entire model for stress period 2. The entries listed on the line labeled "MANAGED FLOW" represent optimal pumping rates to and from WEL-type managed wells only; specified pumping rates at unmanaged WEL-type wells are shown in the entries labeled "WELLS." In contrast, optimal pumping rates to and from MNW-type managed wells are lumped with specified pumping rates at unmanaged MNW-type wells in the entries labeled "MNW2."

(A) Unmanaged flows at WEL-type wells (Qu3 through Qu8)

WELL NO.	LAYER	ROW	COL	STRESS RATE
1	1	3	6	-0.6685E+05
2	1	3	9	-0.6685E+05
3	1	6	9	-0.6685E+05
4	1	6	12	-0.6685E+05
5	1	9	6	-0.6685E+05
6	1	9	9	-0.6685E+05

Managed Flows for Stress Period 2

(B) Managed flows at WEL-type wells (Qm5 and Qm6)

Well no.	Layer	Row	Col	Stress Rate
1	1	9	12	-20000.00000000000
2	1	18	4	0.000000000000000
			Total:	-23115.95311568655

(C) Summary information for both unmanaged (Qu1 and Qu2) and managed (Qm1, Qm2a, Qm2b, and Qm3) MNW-type wells

Summary information for MNW2 wells

WELLID	Totim	Qin	Qout	Qnet	hwell
QU1	10000.0	-20575.2	575.185	-20000.0	165.833
QU2	10000.0	-66850.0	0.00000	-66850.0	148.345
MNWQM1	10000.0	-20000.0	0.00000	-20000.0	149.385
MNWQM2A	10000.0	-2840.87	0.00000	-2840.87	164.231
MNWQM2B	10000.0	-2840.87	0.00000	-2840.87	167.354
MNWQM3	10000.0	-2971.35	0.00000	-2971.35	150.774

(D) Volumetric budget for entire model at end of time step 15 in stress period 2

```
CUMULATIVE VOLUMES      L**3       RATES FOR THIS TIME STEP      L**3/T
     ------------------                 ------------------------

           IN:                                     IN:
           ---                                     ---
             STORAGE =           0.0000                STORAGE =           0.0000
       CONSTANT HEAD =           0.0000          CONSTANT HEAD =           0.0000
               WELLS =           0.0000                  WELLS =           0.0000
              DRAINS =           0.0000                 DRAINS =           0.0000
            RECHARGE = 27300000000.0000               RECHARGE =     2730000.0000
                MNW2 =    73947656.3943                   MNW2 =         575.1853
       MANAGED WELLS =           0.0000          MANAGED WELLS =           0.0000

            TOTAL IN = 27373947656.3943               TOTAL IN =     2730575.1853

          OUT:                                    OUT:
          ----                                    ----
             STORAGE =           0.0000                STORAGE =           0.0000
       CONSTANT HEAD = 14515702278.8720          CONSTANT HEAD =     1407688.8551
               WELLS =  4011000000.0000                  WELLS =      401100.0000
              DRAINS =  8095777840.2290                 DRAINS =      785707.7937
            RECHARGE =           0.0000               RECHARGE =           0.0000
                MNW2 =   651463078.2216                   MNW2 =      116078.2696
       MANAGED WELLS =   100000000.0000          MANAGED WELLS =       20000.0000

           TOTAL OUT = 27373943197.3226              TOTAL OUT =     2730574.9184

            IN - OUT =        4459.0717               IN - OUT =           0.2669

 PERCENT DISCREPANCY =           0.00     PERCENT DISCREPANCY =           0.00
```

Figure 6. Selected output from the MODFLOW **LIST** file for the end of the second stress period and optimal pumping rates for the MNW–SUPPLY sample problem.

Impact of Multi-Node Wells on the GWM Algorithm

The Response Matrix Solution (RMS) Package of GWM–2005 uses the response-matrix method for solving the groundwater-management problem (Ahlfeld and others, 2005). In this approach, a response matrix is constructed by a series of perturbations—one for each decision variable. A flow-process model is first run with all flow-rate decision variables at a user-defined base condition (for example, all pumping turned off). Then the pumping rate at each flow-rate decision variable, in turn, is changed, and the flow-process model is run again. The change in model state (for example, hydraulic head) between the perturbed and base runs provides a means to estimate the response of the system to changes in pumping at the given flow-rate decision variable. The RMS Package assembles the response values into a response matrix and then solves the optimization problem by using this matrix.

The response matrix is a linear approximation of the flow-process model. When the flow-process model itself is linear (for example, for a confined aquifer with no head-dependent boundary conditions), then the approximation is exact. In such cases, the linear programming (LP) solution algorithm is used by GWM–2005. Even if the flow-process model includes nonlinearities, the linear approximation and use of the LP solution algorithm may still be sufficiently accurate to produce an acceptable solution to the optimization problem. Alternatively, an improved solution may be obtained by use of the sequential SLP algorithm (Ahlfeld and others, 2005; Ahlfeld and Baro-Montes, 2008), which iteratively updates the linear approximation.

For problems with nonlinear responses, the accuracy of the response values and the success of the SLP algorithm depend on the degree of nonlinearity. Understanding the characteristics of the nonlinear response is important for the successful use of GWM, which involves the selection of perturbation strategies and solution-algorithm options (for example, LP or SLP approaches). The use of MNW wells affects the GWM calculation of responses in two ways: (1) by the presence of managed or unmanaged MNW wells in the flow-process model, and (2) the perturbation of these wells for the purpose of computing the response matrix.

The solution algorithm of the MNW2 Package incorporates well losses into the calculation of heads and introduces the head in the well as a new dependent variable in the model. At the conclusion of each time step, the algorithm ensures that for each cell associated with a well, the heads in the well and in the cell satisfy the relation described in equation 1.

As noted previously, in addition to satisfying equation 1, MNW also requires that the sum of flow rates to or from each cell associated with the well equal the total net flow rate from the well (Q_{net}):

$$Q_{net} = \sum_{i=1}^{n} Q_i \, , \tag{5}$$

where

Q_i is the flow rate to or from the well from cell i, and
n is the number of cells to which the well is connected.

For each MNW well, the MNW algorithm can be thought of as adding $n+1$ variables that need to be solved by the flow-process model; these are the n cell flows and the head at the well. The resulting system of equations will be linear if the value of the variable C in equation 1 is set to zero, and all other aspects of the model are linear (that is, for a confined aquifer with no head-dependent boundary conditions). If the system of equations is linear, then a response matrix computed from the flow-process model will be an exact approximation of the flow process.

To demonstrate an analysis of the nonlinearity produced by including MNW wells, a test case using the MNW–SUPPLY sample problem is examined. For this demonstration, groundwater-level (head) responses for one of the MNW-type wells ($Qm1$, shown in fig. 1A) and for the two vertically adjacent cells that span the open interval of the well are examined. This approach to the demonstration is anticipated to give the most severe nonlinear responses. The sample problem is converted to a confined condition, and pumping rates are such that the drain cells never dry, so that except for the MNW well, the response of head to pumping is linear. Transmissivities for layers 1 and 2 are identical and equal to 15,000 square feet per day (ft^2/d). The saturated thickness is 250 ft for layer 1 and 50 ft for layer 2. For the simulations described below, constant pumping rates are specified at the well for all five stress periods.

In the first set of simulations, LOSSTYPE at the well is selected to be SKIN: the radius of the well (Rw) in each layer is specified to be 0.5 ft, the radius of the skin (Rskin) in each layer is 1.0 ft, and the hydraulic conductivity of the skin (Kskin) is 24.6 ft/d in layer 1 and 122.8 ft/d in layer 2. For these simulated conditions, variable A in equation 1 is equal to 7.37×10^{-5} d/ft^2 for both layers, variable B is 2.26×10^{-5} d/ft^2 in layer 1 and 1.06×10^{-5} d/ft^2 in layer 2, and variable C is zero for both layers. Groundwater-level responses in the well and the two cells associated with the well at the end of the fourth stress period are shown in figure 7A. For pumping rates below about 1,180,000 ft^3/d, the responses of all three heads are linear, as expected; for pumping rates above this value, the responses are again linear, but the slopes of the lines are different. The slopes change when

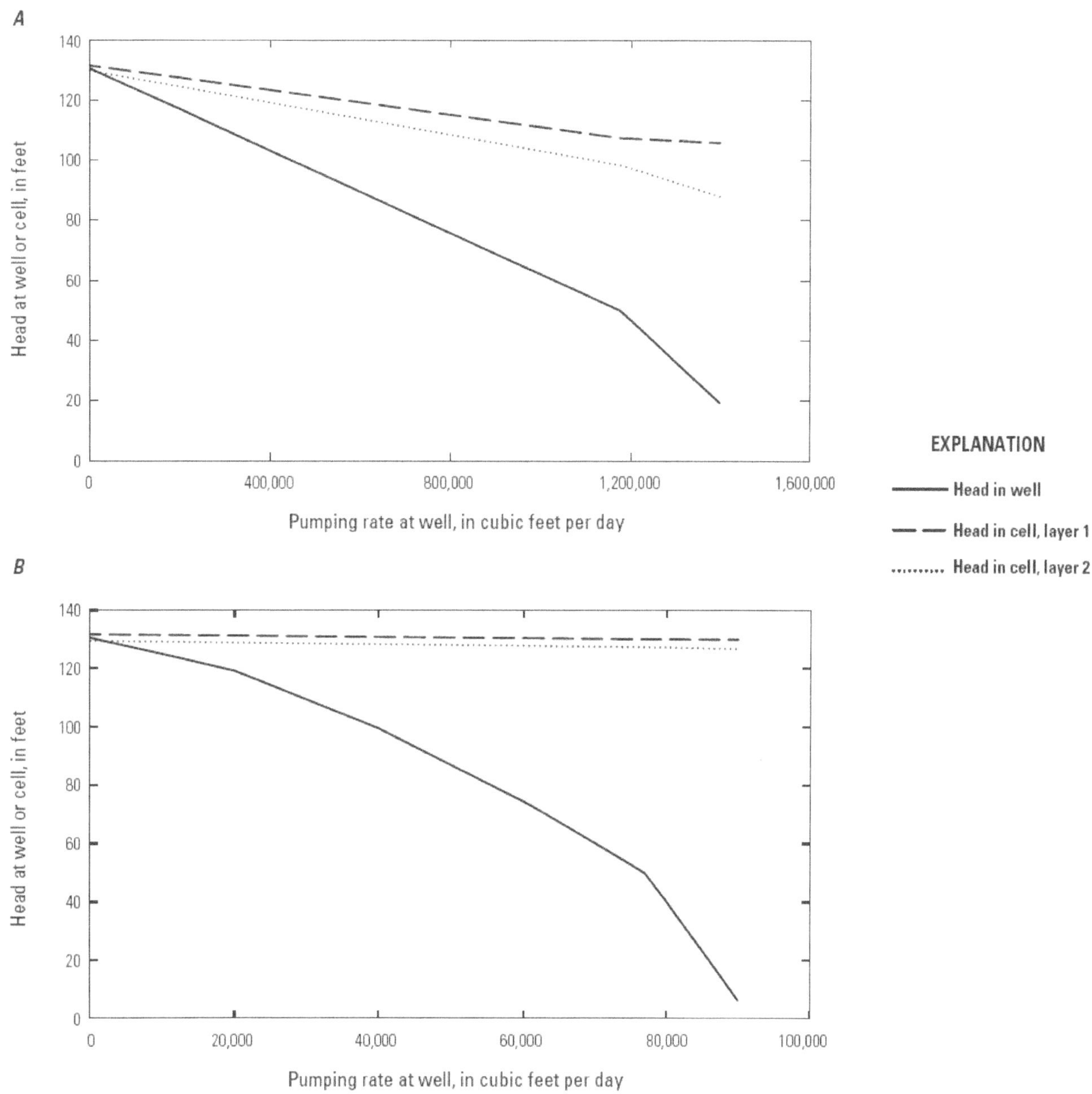

Figure 7. Groundwater-head responses at the end of the fourth stress period to pumping at well *Qm1* for MNW–SUPPLY sample problem with input-variable LOSSTYPE equal to *A*, SKIN and *B*, GENERAL. Well location shown on figure 1*A*.

the heads in the well reach the bottom of the first layer; at this point, the MNW algorithm converts the withdrawals in this layer to a third-type boundary condition (head-dependent flux), producing a nonlinear effect.

In a second set of simulations, LOSSTYPE at the well is selected to be GENERAL. The values for Rw, A , and B are the same as they were for the SKIN LOSSTYPE simulations, but now variables C and P are specified to be 1.0×10^{-5} $d^{1.5}/ft^{3.5}$ and 1.5 for both layers. Results for these simulations are shown in figure 7B. Again, the head response changes when the head in the well reaches a value of 50 ft at a pumping rate of about 76,800 ft^3/d; this rate is much lower than for the SKIN LOSSTYPE because the well losses are so much higher. At pumping rates below this value, the response is nonlinear as a result of the last term in equation 1, which is nonlinear and now active. For this hypothetical example, the head losses generated by selection of the GENERAL LOSSTYPE are intentionally very large to demonstrate the possible nonlinear responses. For example, at a pumping rate of 60,000 ft^3/d, the head difference between head in the top cell and in the well is about 4 ft for the SKIN LOSSTYPE case; at the same pumping rate and with the C term added as shown in figure 7B, the head difference is about 56 ft. This comparison indicates that the nonlinear C term in equation 2 dominates the response for the example depicted in figure 7B.

This sample problem shows that when the value of C in equation 1 is set to zero (that is, for all cases except LOSSTYPE = GENERAL), the presence of MNW wells adds nonlinearity only when the head in the well falls below (or rises above) the elevation of the layer bottom (or top). This behavior is similar to that with head-dependent boundary conditions such as drain cells or stream cells that disconnect from the aquifer as heads drop. Even when the GENERAL LOSSTYPE is used, the value of C must be quite large for the nonlinear response to be substantial.

At the conclusion of each perturbation run, GWM–2005 checks the results from the flow process to determine if they are acceptable. If they are not acceptable, the perturbation value is changed and the run repeated. One of the checks determines if the cell in which the managed WEL-type well is located has dewatered. A check similar to this is implemented for MNW-type wells. The actual pumping rate determined by MNW (Q_{net}) is compared with the pumping rate given to MNW by GWM. If they differ, then the perturbation run is assumed to have been unsuccessful and is repeated with a lower pumping rate. A second check by GWM–2005 is done to determine if the heads at head-constraint locations or locations defined as head-type state variables have dewatered (=HDRY) at the conclusion of a run. This check has been extended to the heads in managed MNW wells to determine if they are constrained or have been defined as state variables. The intention of these post-perturbation checks is to identify perturbation-run results that will not produce linear or near-linear system responses. The user is advised to carefully select the GWM perturbation parameters so that they have a realistic range (that is, a fraction of the maximum pumping rate for the well). Moreover, users should be aware that under some conditions, the unmanaged internal flows within a borehole (as represented by the values *Qin* and *Qout*) may be quite large. Such conditions can include, for example, substantial borehole flow between model layers. The GWM solution algorithm should be able to handle these types of large internal flows, but the large flows may cause unexpected flow conditions in the simulated aquifer. The user is encouraged to check all of the output from a GWM run, including results reported by MODFLOW–2005 for the simulated aquifer and MNW wells, to be sure that the managed and unmanaged aspects of the system behave as intended.

Summary

This report describes the use of multi-node wells (MNW) in groundwater-management problems that are formulated and solved with the GWM–2005 Process for MODFLOW–2005. This new capability expands the range of applicability of GWM–2005 by allowing MNW-type wells to be incorporated into a management formulation in two ways: (1) as flow-rate decision variables, in which the flow rate at each managed MNW-type well is determined as part of the optimization process; and (2) as sites for management of water levels, at which the water level at an MNW-type managed well is constrained by a head-type constraint or managed by use of head-type state variables. Both management options can be simulated simultaneously at an MNW-type well; in other words, it is possible to manage water levels at an MNW-type flow-rate decision variable. The use of MNW-type wells as management variables was tested successfully with a representative sample problem. The sample problem demonstrates that MNW-type wells add nonlinearities to a management formulation that are similar to those caused by other head-dependent boundary conditions such as drain or stream cells, and that the most substantial nonlinearities are caused by turbulent flow near a multi-node well.

Acknowledgments

The authors thank Kevin Mulligan of the University of Massachusetts Amherst, who conducted extensive testing of the new code and helped to create the sample problem. The authors also thank Mary Ashman, Leonard Konikow, Derek Ryter, and Brian Wagner of the U.S. Geological Survey for their helpful comments on an earlier draft of this report.

References Cited

Ahlfeld, D.P., Baker, K.M., and Barlow, P.M., 2009, GWM-2005—A Groundwater-Management Process for MODFLOW-2005 with Local Grid Refinement (LGR) capability: U.S. Geological Survey Techniques and Methods, book 6, chap. A33, 65 p.

Ahlfeld, D.P., Barlow, P.M., and Baker, K.M., 2011, Documentation for the State Variables Package for the Groundwater-Management Process of MODFLOW-2005 (GWM-2005): U.S. Geological Survey Techniques and Methods, book 6, chap. A36, 45 p.

Ahlfeld, D.P., Barlow, P.M., and Mulligan, A.E., 2005, GWM—A Ground-Water Management Process for the U.S. Geological Survey modular ground-water model (MODFLOW-2000): U.S. Geological Survey Open-File Report 2005–1072, 124 p.

Ahlfeld, D.P., and Baro-Montes, Gemma, 2008, Solving unconfined groundwater flow management problems with successive linear programming: Journal of Water Resources Planning and Management, v. 134, no. 5, p. 404–412.

Halford, K.J., and Hanson, R.T., 2002, User guide for the drawdown-limited, multi-node well (MNW) package for the U.S. Geological Survey's modular three-dimensional finite-difference ground-water flow model, versions MODFLOW-96 and MODFLOW-2000: U.S. Geological Survey Open-File Report 02–293, 33 p.

Harbaugh, A.W., 2005, MODFLOW-2005, The U.S. Geological Survey modular ground-water model—The Ground-Water Flow Process: U.S. Geological Survey Techniques and Methods, book 6, chap. A16 [variously paged].

Konikow, L.F., Hornberger, G.Z., Halford, K.J., and Hanson, R.T., 2009, Revised multi-node well (MNW2) package for MODFLOW ground-water flow model: U.S. Geological Survey Techniques and Methods, book 6, chap. A30, 67 p.

Appendix 1. Input Instructions for Use of MNW-Type Wells in a GWM Formulation

Contents

Input Instructions for Adding MNW-Type Wells to a GWM Formulation 20
MODFLOW **NAME** File 20
Decision Variables (**DECVAR**) File 20
State Variables (**STAVAR**) File 24
Head Constraints (**HEDCON**) File 25
References Cited 26

Input Instructions for Adding MNW-Type Wells to a GWM Formulation

This appendix describes modifications to the Groundwater-Management (GWM) Process data-input files that are necessary when multi-node wells (MNW) simulated by the MNW2 Package (Konikow and others, 2009) are included in a GWM management problem. GWM-2005 allows MNW wells to be either unmanaged or managed. Input data for unmanaged wells are defined in the MNW2 Package input file, as described by Konikow and others (2009). When managed wells are included in a GWM formulation, they must be defined as flow-rate decision variables in the GWM decision-variables (**DECVAR**) input file; upper and lower bounds for each managed well also must be defined in the decision-variable constraints (**VARCON**) file (Ahlfeld and others, 2005). In addition, if water-level constraints are defined at MNW wells in a management formulation, then these constraints can be defined in the head-constraints (**HEDCON**) file (Ahlfeld and others, 2005). Water levels at MNW wells also can be defined as head-type state variables with information defined in the GWM state-variables (**STAVAR**) file (Ahlfeld and others, 2011) and possibly also in the summation-constraints (**SUMCON**) file (Ahlfeld and others, 2005). The sections that follow provide line-by-line input instructions for entry of data about MNW wells in the **DECVAR**, **STAVAR**, and **HEDCON** files; no special data-input instructions are needed for the **VARCON** or **SUMCON** files for the use of managed MNW-type flow-rate decision variables. Up-to-date data-input instructions for all GWM input files are provided with each release of the GWM software at the Internet address shown in the Preface of this report.

MODFLOW NAME File

The implementation for communicating information about both managed and unmanaged MNW wells to the MODFLOW groundwater-flow process involves writing a temporary MNW2 Package input file. GWM–2005 reads both the original MNW2 input file, defined by the user and listed in the MODFLOW **NAME** file, and the **DECVAR** file. GWM–2005 requires that an MNW2 input file be listed in the **NAME** file if the **DECVAR** file includes any MNW-type decision variables. If there are no unmanaged MNW wells, then the MNW2 input file should be specified with the MNW2 input variable MNWMAX, which is used to define the number of MNW wells, set to zero. GWM–2005 uses the information from both files to write a new file in a format that can be read by the MNW2 Package as a standard MNW2 input file. GWM–2005 uses the dedicated file name "GWM_MNW_Combined_Input.txt" for this purpose. The file is created for each run of the MODFLOW groundwater-flow process, each time with a different set of specified flow rates for the managed MNW wells. GWM–2005 reads the user-supplied MNW2 input file only once; it stores information needed for future writing of the combined file. The original user-supplied MNW2 input file is not affected by this process. At the conclusion of a GWM–2005 run, the final combined file, which includes the optimal pumping rates for a successful run, remains in the execution directory and can be discarded.

Decision Variables (DECVAR) File

The decision-variables (**DECVAR**) file is used to define the three types of decision variables that may be used in a GWM management model: flow-rate, external, and binary. The only modifications that need to be made to the **DECVAR** file for use of MNW-type flow-rate decision variables are to input items 2 and 3, as shown below. Only the input variables for items 2 and 3 are defined below; users of GWM should refer to the input instructions distributed with the software for definitions of all variables defined in the **DECVAR** file. Moreover, users should be aware that the information specified in a **DECVAR** file for each MNW-type flow-rate variable differs in some respects from the information that is specified for an unmanaged MNW-type well in an **MNW2** input file.

Descriptions of input items 2 and 3a through 3e follow; descriptions for each of the input variables are provided after the input-item definitions. An example **DECVAR** file for the MNW–SUPPLY sample problem is shown in figure 2 of the report.

Item 2. NFVAR NEVAR NBVAR
Item 3a. The following records are read for each of the NFVAR flow-rate decision variables:
FVNAME NC LAY ROW COL FTYPE FSTAT WSP

Options for specifying NC, LAY, ROW, and COL: Input variable NC is an integer variable that describes the type and number of wells that are associated with a flow-rate decision variable. In previous versions of GWM, NC always referred to WEL-type wells. With the addition of the MNW functionality, NC can refer to either a WEL-type or MNW-type well: if NC is specified as a positive integer, then the flow-rate decision variable consists of one or more WEL-type wells; if NC is specified as a negative integer, then the flow-rate decision variable consists of one or more MNW-type wells. A flow-rate decision variable cannot consist of a mix of WEL-type and MNW-type wells; it must be either all WEL-type or all MNW-type wells. Also, NC cannot be specified as zero (if it is, GWM will write an error message and stop execution).

Four options are provided for specifying NC. The first two options are used for WEL-type and the second two options for MNW-type flow-rate decision variables. Depending on the value specified for NC, the user must optionally complete items 3b to 3e, which are described below.

- If NC = 1, then the flow-rate decision variable consists of a single WEL-type withdrawal or injection well at a single model cell defined by LAY, ROW, COL. The user must complete item 3a, skip items 3b through 3e, and then proceed to the next flow-rate decision variable (that is, to item 3a); or, if this is the last variable to be defined, proceed to items 4 and 5.
- If NC ≥ 2, then the flow-rate decision variable consists of a total of NC WEL-type withdrawal or injection wells at model cells that must be defined in item 3b-1 below. The flow rate calculated for this decision variable is distributed over the NC cells in proportion to the values specified by RATIO for each cell in item 3b-1. Item 3b-1 is read NC times. Values must be specified for variables LAY, ROW, and COL in item 3a, but the specified values are ignored by GWM (for example, the user could specify values of zero for each of the three variables). The user must complete items 3a and 3b-1, skip items 3b-2 through 3e, and then proceed to the next flow-rate decision variable (that is, to item 3a); or, if this is the last variable to be defined, proceed to items 4 and 5.
- If NC = -1, then the flow-rate decision variable consists of a single MNW-type withdrawal or injection well. Values must be specified for variables LAY, ROW, and COL in item 3a, but the specified values are ignored by GWM (for example, the user could specify values of zero for each of the three variables). The user must complete items 3b-2 and 3c through 3e (and skip items 3b-1 and 3b-3) and then proceed to the next flow-rate decision variable (that is, to item 3a); or, if this is the last variable to be defined, proceed to items 4 and 5.
- If NC ≤ -2, then the flow-rate decision variable consists of a total of the absolute value of NC MNW-type withdrawal or injection wells. The flow rate calculated for this decision variable is distributed over the absolute value of NC multi-node wells in proportion to the values specified by RATIO for each well. Values must be specified for variables LAY, ROW, and COL in item 3a, but the specified values are ignored by GWM (for example, the user could specify values of zero for each of the three variables). The user must complete items 3b-3 through 3e (and skip items 3b-1 and 3b-2) for each of the multi-node wells associated with this decision variable, and then proceed to the next flow-rate decision variable (that is, to item 3a); or, if this is the last variable to be defined, proceed to items 4 and 5.

Item 3b. One of the following input items must be completed if NC is not equal to 1:

3b-1. If NC ≥ 2, repeat the following item NC times: RATIO LAY ROW COL
3b-2. If NC = -1: WELLID NNODES
3b-3. If NC ≤ -2: RATIO WELLID NNODES

Item 3c. LOSSTYPE PPFLAG

Item 3d. If LOSSTYPE = NONE, then skip item 3d, and proceed to item 3e. If LOSSTYPE ≠ NONE, then include item 3d by specifying values for one or more of the following variables depending on the type of LOSSTYPE specified in item 3c:

{Rw Rskin Kskin B C P CWC}

If LOSSTYPE = THIEM, specify Rw.
If LOSSTYPE = SKIN, specify Rw, Rskin, and Kskin.
If LOSSTYPE = GENERAL, specify Rw, B, C, and P.
If LOSSTYPE = SPECIFYcwc, then specify CWC.

Any of the variables in item 3d can be assumed to be constant for the entire length of the open interval of the well (in which case appropriate values are simply specified here in item 3d), or they can be assumed to vary along the length of the open interval of the well (in which case any negative value should be specified here in item 3d and the actual values then specified for each node or open interval in item 3e). For example, if Rw is specified as -1, then Rw is assumed to vary along the length of the well, and a real value of Rw must be defined for each node (or open interval) of this well in item 3e.

Item 3e. The user must enter location information for each multi-node well in item 3e. For each node (or open interval) of this well, the user must use input format 3e-1 or 3e-2, depending on the value of NNODES. In either case, item 3e must consist of a total number of records equal to the absolute value of NNODES. If the relevant LOSSTYPE variables were set to negative values in item 3d, then they vary in value among nodes (or open intervals, if NNODES < 0) and should be defined here in item 3e according to the definitions given under item 3d. The values specified here, if any, are those that were set to a negative value in item 3d.

For NNODES > 0, enter one line of input data for each of the NNODES model cells (nodes) for the current well:

3e-1. LAY ROW COL {Rw Rskin Kskin B C P CWC PP}

The data list of nodes defining the multi-node well must be constructed and ordered so that the first node listed represents the node closest to the wellhead, the last node listed represents the node farthest from the wellhead, and all nodes are listed in sequential order from the top to the bottom of the well. A particular node in the grid can be associated with more than one multi-node well.

For NNODES < 0, the absolute value of NNODES indicates how many open intervals are to be defined and so must correspond exactly to the number of records in item 3e-2 for this well:

3e-2. Ztop Zbotm ROW COL {Rw Rskin Kskin B C P CWC PP}

The data list of intervals defining the multi-node well must be constructed and ordered so that the first interval listed represents the shallowest one, the last interval listed represents the deepest one, and all intervals are listed in sequential order from the top to the bottom of the well. If an interval partially or fully intersects a model layer, then a node will be defined in that cell. If more than one open interval intersects a particular layer, then a length-weighted average of the cell-to-well conductances will be used to define the well-node characteristics; for purposes of calculating the effects of partial penetration, the cumulative length of the well screens will be assumed to be centered vertically within the thickness of the cell. If the well is a single-node well by definition of LOSSTYPE = NONE, and the defined open interval straddles more than one model layer, then the well will be associated with the cell that contains the center of the open interval. If the specified elevations indicate multiple well screens or open intervals within a single model layer, then the model will assign a single composite well screen for the node based on the ratio of the screen length to layer thickness (see discussion in Konikow and others, 2009). However, if CWC values are specified by the user, then these values are assumed to be already appropriate for the actual length of the screen and are not adjusted by this ratio.

Variables for input items 2 and 3 are defined below; the remaining variables are defined in the input instructions for GWM distributed with the software.

NFVAR—is an integer variable equal to the total number of flow-rate decision variables simulated by either the WEL- or MNW-type functionalities. NFVAR must be greater than 0. Only one flow-rate decision variable can be defined for a particular combination of well locations and stress periods associated with the decision variable, with the exceptions that both a withdrawal variable (FTYPE = W) and an injection variable (FTYPE = I) can be defined simultaneously for the same combination of well locations and stress periods.

NEVAR—is an integer variable equal to the number of external decision variables. NEVAR must be greater than or equal to 0.

NBVAR—is an integer variable equal to the number of binary variables. NBVAR must be greater than or equal to 0. If NBVAR is 0, binary variables are not included in the management formulation.

FVNAME—is a character variable up to 10 characters long that is a unique name designated for the flow-rate decision variable. Each name must be unique (that is, the same name cannot be used for more than one variable or in more than one model when the local-grid refinement capability of GWM is used). No spaces are allowed in the name. The end of the name is designated by a blank space.

NC—is an integer variable that describes the type and number of wells that are associated with a flow-rate decision variable. See the discussion above to determine the correct value to specify for NC.

LAY, ROW, and COL—are integer variables equal to the layer, row, and column number of the model cell or multi-node well to which the flow for decision-variable FVNAME will be assigned. See the discussion above to determine the correct values to specify for LAY, ROW, and COL.

FTYPE—is a character variable that indicates whether the decision variable is a withdrawal or injection site. If FTYPE is W, the site is used for withdrawal; if FYPTE is I, the site is used for injection. If either withdrawal or injection is allowed at the site, two decision variables must be defined for the site: one for withdrawal (that is, with FTYPE = W) and one for injection (FTYPE = I).

FSTAT—is a character variable that indicates whether the decision variable will be considered in the management problem. If FSTAT is Y, the decision variable is available; if FSTAT is N, the decision variable is unavailable. If the decision variable is unavailable, then no withdrawal or injection will be calculated for the decision-variable location. (In most cases, FSTAT will be set to Y.)

WSP—is a character string up to 120 characters long that indicates the stress periods associated with decision variable FVNAME. A single flow rate will be determined by GWM for all the stress periods included in WSP. The string must not contain any blank spaces. Multiple stress periods are indicated by colons (:), hyphens (-), or combinations of colons and hyphens. For example,

1 indicates that stress period 1 is the only stress period associated with the decision variable,
1:3 indicates that the flow rate is the same for stress periods 1 and 3 (but not 2),
1–12 indicates that the flow rate is the same for stress periods 1 through 12, and
2:5–7:12 indicates that the flow rate is the same for stress periods 2, 5, 6, 7, and 12.

RATIO—is a real variable. RATIO is the fraction of the total flow rate for decision variable FVNAME that is distributed to each individual cell or multi-node well associated with decision variable FVNAME. The sum of the RATIO values must equal 1.0 for all of the NC cells or multi-node wells specified for FVNAME; if the sum does not equal 1.0, GWM calculates the fraction for each cell or multi-node well by dividing the RATIO value specified for each cell or multi-node well by the sum of the RATIO values specified for all cells or multi-node wells associated with FVNAME.

WELLID—is a character variable that is a unique alphanumeric identification label for each multi-node well. The text string is limited to 20 alphanumeric characters. If the name of the well includes spaces, then enclose the name in quotation marks.

NNODES—is an integer variable equal to the number of cells (nodes) associated with this multi-node well. NNODES normally is specified to be > 0, but for the case of a vertical borehole, setting NNODES < 0 will allow the user to specify the elevations of the tops and bottoms of well screens or open intervals (rather than grid layer numbers). The absolute value of NNODES equals the number of open intervals (or well screens) to be specified in dataset 3e. If this option is used, then the model will compute the layers that contain the open intervals, the lengths of the open intervals, and the relative vertical positions of the open intervals within each model layer (for example, see fig. 14 and related discussion in Konikow and others, 2009).

LOSSTYPE—is a character variable used to determine the user-specified model for well loss (equation 1). Available options are

NONE—there are no well corrections, and the head in the well is assumed to equal the head in the cell. This option ($h_{well} = h_n$) is valid only for a single-node well (NNODES = 1).

THIEM—this option allows for only the cell-to-well correction based on the Thiem (1906) equation; the head in the well is determined from equation 1 as $h_{well} = h_n + AQ_n$, and the model computes A on the basis of the user-specified well radius (Rw, defined below). Coefficients B and C in equation 1 are automatically set to zero.

SKIN—this option allows for formation damage or skin corrections at the well: $h_{well} = h_n + AQ_n + BQ_n$ from equation 1, where A is determined by the model from the value specified for the variable Rw (defined below), and B is determined by the model from the values specified for variables Rskin and Kskin (defined below).

GENERAL—head loss is defined with coefficients A, B, and C and power exponent P ($h_{well} = h_n + AQ_n + BQ_n + CQ_n^P$). Coefficient A is determined by the model from the value specified for the variable Rw (defined below). User must also specify variables B, C, and P, which are defined below. A value of P = 2.0 is suggested if no other data are available (the model allows $1.0 \le P \le 3.5$). Entering a value of $C = 0$ will result in a linear model in which the value of B is entered directly (rather than entering properties of the skin, as with the SKIN option).

SPECIFYcwc—the user specifies an effective conductance value (equivalent to the combined effects of the A, B, and C well-loss coefficients expressed in equation 15 in Konikow and others, 2009) between the well and the cell representing the aquifer through variable CWC (defined below). If there are multiple screens within the grid cell, or if partial-penetration corrections are to be made, then the effective value of CWC for the node may be further adjusted automatically by MNW2.

PPFLAG—is an integer variable that determines whether the calculated head in the well will be corrected for the effect of partial penetration of the well screen in the cell. If PPFLAG = 0, then the head in the well will not be adjusted for the effects of partial penetration. If PPFLAG > 0, then the head in the well will be adjusted for the effects of partial penetration if the section of well containing the well screen is vertical (as indicated by identical row-column locations in the grid). If NNODES < 0 (that is, the open intervals of the well are defined by top and bottom elevations), then the model will automatically calculate the fraction of penetration for each node and the relative vertical position of the well screen. If NNODES > 0, then the fraction of penetration for each node must be defined in dataset 3e, and the well screen will be assumed to be centered vertically within the thickness of the cell (except if the well is in the uppermost unconfined model layer, in which case the bottom of the well screen will be assumed to be aligned with the bottom boundary of the cell, and the assumed length of well screen will be based on the initial head in that cell).

Rw—is a real variable equal to the radius of the well (L).

Rskin—is a real variable equal to the radius to the outer limit of the skin (L).

Kskin—is a real variable equal to the hydraulic conductivity of the skin (L/T).

B—is a real variable equal to the linear well-loss coefficient in equation 1 (T/L^2).

C—is a real variable equal to the nonlinear well-loss coefficient in equation 1 ($T^P/L^{(3P-1)}$).

P—is a real variable equal to the power (exponent) of the nonlinear discharge component of well loss in equation 1 (dimensionless).

CWC—is a real variable equal to the cell-to-well hydraulic conductance in equation 15 in Konikow and others (2009) (L^2/T).

PP—is a real variable equal to the fraction of partial penetration for this cell (see PPFLAG). Only specify if PPFLAG > 0 and NNODES > 0.

Ztop, Zbotm—are real variables equal to the top and bottom elevations of the open intervals (or screened intervals) of a vertical well.

State Variables (STAVAR) File

GWM allows heads at managed MNW-type wells to be defined as head-type state variables by use of a **STAVAR** input file. The MNW wells must be defined as flow-rate decision variables in the **DECVAR** file for the management formulation. Head-type state variables defined at MNW-type wells are included with other head-type state variables under the NHVAR input variable in item 2 of the **STAVAR** file. Although the **STAVAR** file includes a total of six input items, only items 2 and 3 need to be modified to include head-type state variables at MNW wells; users of GWM should refer to the input instructions distributed with the software for a complete set of instructions for all variables defined in the **STAVAR** file. There are two options for item 3: head-type state variables specified at model cells are defined in item 3a, whereas head-type state variables specified for MNW wells are defined in item 3b. The variable MFLG in item 3b is a flag that must be set to zero and indicates to GWM that the line of input data is formatted as item 3b (head-type state variable at an MNW well). The variable WELLID is the name of an MNW well that has been defined in the **DECVAR** file for the management formulation; GWM uses this name to find the location in the flow-process model where the constraint will be defined.

Descriptions of input items 2, 3a, and 3b follow; a description of each of the input variables is provided after the input-item definitions. An example use of head-type state variables at managed MNW wells is provided in the MNW–SUPPLY sample problem described in the report.

2. NHVAR NRVAR NSVAR NDVAR

3a. The following record is read for each head-type state variable specified at a model cell:

SVNAME LAY ROW COL SVSP

3b. The following record is read for each head-type state variable specified at an MNW-type well that has been defined in the **DECVAR** file:

SVNAME MFLG WELLID SVSP

Variables for input items 2 and 3 are defined below; the remaining variables are defined in the input instructions for GWM distributed with the software.

NHVAR—is an integer variable equal to the total number of head-type state variables specified at cells or MNW-type managed wells. NHVAR must be greater than or equal to 0.

NRVAR—is an integer variable equal to the number of streamflow-type state variables. NRVAR must be greater than or equal to 0.

NSVAR—is an integer variable equal to the number of storage-type state variables. NSVAR must be greater than or equal to 0.

NDVAR—is an integer variable equal to the number of drain state variables. NDVAR must be greater than or equal to 0.

SVNAME—is a character variable up to 10 characters long that is a unique name designated for the state variable. The names for the head, streamflow, and drain state variables must be unique (that is, the same name cannot be used for more than one variable, or in more than one model when the local-grid refinement capability of GWM is used). For storage state variables applied to multimodel problems (that is, those using local grid refinement), the same name may appear in more than one **STAVAR** file to define a storage state variable that extends over multiple grids. However, in a given **STAVAR** file on one grid of a multimodel problem, the state variable name must be unique. No spaces are allowed in the name. The end of the name is designated by a blank space.

LAY, ROW, and COL—are integer variables equal to the layer, row, and column numbers of the model cells in which the head-type or drain-type state variables are located. For drain state variables, the LAY, ROW, and COL numbers must correspond to a valid drain location as specified in the MODFLOW DRN input file.

SVSP—is an integer variable that indicates the stress period during which the head, streamflow, or drain flow-rate state variables are to be evaluated. To evaluate these state variables for multiple stress periods, define multiple state variables.

MFLG—is an integer variable that must be specified as 0. The variable is a flag that indicates to GWM that a head-type state variable is being defined for an MNW well.

WELLID—is a character variable that is the unique alphanumeric identification of an MNW-type well defined in the **DECVAR** file. The text string is limited to 20 alphanumeric characters. If the name of the well includes spaces, then enclose the name in quotation marks.

Head Constraints (HEDCON) File

GWM allows head-bound constraints to be imposed on the head in a managed MNW-type well by use of a **HEDCON** input file. Heads in MNW wells cannot be used in drawdown, head-difference, or gradient constraints. The MNW wells must be defined as flow-rate decision variables in the **DECVAR** file for the management formulation. Head-bound constraints on heads at MNW-type wells are included with other head-bound constraints under the NHB input variable in item 2 of the **HEDCON** file. Although the **HEDCON** file includes a total of six input items, only items 2 and 3 need to be modified to include head-bound constraints at MNW wells; users of GWM should refer to the input instructions distributed with the software for definitions of all variables defined in the **HEDCON** file. Item 3 includes two options: head-bound constraints specified at model cells are defined in item 3a, whereas head-bound constraints specified at MNW wells are defined in item 3b. The variable MFLG in item 3b is a flag that must be set to zero and indicates to GWM that the line of input data is formatted as item 3b (head-bound constraint at an MNW well). The variable WELLID is the name of an MNW well that has been defined in the **DECVAR** file for the management formulation; GWM uses this name to find the location in the flow-process model where the constraint will be imposed.

Descriptions of input items 2, 3a, and 3b follow; the description for each of the input variables is provided after the input-item definitions. An example use of head-bound constraints at managed MNW wells is provided in the MNW–SUPPLY sample problem described in the report.

2. NHB NDD NDF NGD

3a. The following record is read for each head constraint calculated at a model cell:
HBNAME LAYH ROWH COLH TYPH BND NSP

3b. The following record is read for each head constraint calculated at an MNW-type well that has been defined in the **DECVAR** file: HBNAME MFLG WELLID TYPH BND NSP

Variables for input items 2 and 3 are defined below; the remaining variables are defined in the input instructions for GWM distributed with the software.

NHB—is an integer variable equal to the total number of head-bound constraints defined in items 3a and 3b that need to be satisfied in the management model.

NDD—is an integer variable equal to the number of drawdown constraints that need to be satisfied in the management model.

NDF—is an integer variable equal to the number of head-difference constraints that need to be satisfied in the management model.

NGD—is an integer variable equal to the number of gradient constraints that need to be satisfied in the management model.

HBNAME—is a character variable up to 10 characters long that is a unique name designated for the head-bound constraint. No spaces are allowed in the name. The end of the name is designated by a blank space.

LAYH, ROWH, and COLH—are integer variables equal to the layer, row, and column numbers of the model cell in which the head-bound constraint is located.

TYPH—is a character variable used to specify the type of head bound. Two options are allowed:

LE indicates that the head calculated by the model must be less than or equal to the value specified by BND, and
GE indicates that the head calculated by the model must be greater than or equal to the value specified by BND.

BND—is a real variable equal to the specified upper or lower bound on the head at the model cell or multi-node well at the end of the stress period.

NSP—is an integer variable that indicates the stress period during which the constraint is imposed. If the constraint is imposed over multiple stress periods, then a separate record must be provided for each stress period.

MFLG—is an integer variable that must be specified as 0. The variable is a flag that indicates to GWM that a head-bound constraint is being defined at an MNW well.

WELLID—is a character variable that is the unique alphanumeric identification of an MNW-type well defined in the **DECVAR** file. The text string is limited to 20 alphanumeric characters. If the name of the well includes spaces, then enclose the name in quotation marks.

References Cited

Ahlfeld, D.P., Barlow, P.M., and Baker, K.M., 2011, Documentation for the State Variables Package for the Groundwater-Management Process of MODFLOW-2005 (GWM-2005): U.S. Geological Survey Techniques and Methods, book 6, chap. A36, 45 p.

Ahlfeld, D.P., Barlow, P.M., and Mulligan, A.E., 2005, GWM—A Ground-Water Management Process for the U.S. Geological Survey modular ground-water model (MODFLOW-2000): U.S. Geological Survey Open-File Report 2005–1072, 124 p.

Konikow, L.F., Hornberger, G.Z., Halford, K.J., and Hanson, R.T., 2009, Revised multi-node well (MNW2) package for MODFLOW ground-water flow model: U.S. Geological Survey Techniques and Methods, book 6, chap. A30, 67 p.

www.ingramcontent.com/pod-product-compliance
Lightning Source LLC
Chambersburg PA
CBHW081414170526
45166CB00010B/3335
9781500296766